FACES AND VASES!

OPTICAL ILLUSIONS

by Spencer Brinker

Minneapolis, Minnesota

Credits
Cover and Title Page, © Kay Cee Lens and Footages/Shutterstock, © Olena.07/Shutterstock, and © Quang Ho/Shutterstock; 4, © Tadeusz Wejkszo/Shutterstock, © Markus Mainka/Shutterstock, and © Dionisvera/Shutterstock; 5, © istanbulimage/iStock, © Givaga /iStock, © camaralenta/iStock, © Pavlo S/Shutterstock, and © metamorworks/Shutterstock; 6, © PeterHermesFurian/iStock; 7, My Wife and My Mother-in-Law, Courtesy Library of Congress/; 8, © Anneka/Shutterstock, © Kuznetsov Alexey/Shutterstock, and Duck and Rabbit/Public Domain; 9, © Yellowj/Shutterstock; 10, © Sofia Kozlova/Shutterstock; andCourtesy Cooper Hewitt, Smithonian Design Museum; 12, © Ranta Images/Shutterstock; 12-13, © ExpressVectors/Shutterstock; 16, © Goldilock Project/Shutterstock; and © jmsilva/iStock; 17, © Hernan E. Schmidt/Shutterstock, © DG Stock/Shutterstock, © Nathaneal Largent/Shutterstock, © Rashevskyi Viacheslav/Shutterstock, and © Vertumnus, by Giuseppe; 18, All Is Vanity by Charles Allan Gilbert/Public Domain; 18-19, © Thomas Quack/Shutterstock; 19, © The Ambassadors by Hans Holbein/Public Domain; 20, © Franzi/Shutterstock; andPuzzle / the end by Peter Newell/Public Domain; 21, © PeopleImages/iStock; 22, © Epps_Taniam/Shutterstock; and 23, © Simona Sirio/iStock.

Bearport Publishing Company Product Development Team
President: Jen Jenson; Director of Product Development: Spencer Brinker; Senior Editor: Allison Juda; Associate Editor: Charly Haley; Senior Designer: Colin O'Dea; Associate Designer: Elena Klinkner; Editorial Assistant: Naomi Reich.

Library of Congress Cataloging-in-Publication Data is available at www.loc.gov or upon request from the publisher.

ISBN: 978-1-63691-499-2 (hardcover)
ISBN: 978-1-63691-504-3 (ebook)

Copyright © 2022 Bearport Publishing Company. All rights reserved. No part of this publication may be reproduced in whole or in part, stored in any retrieval system, or transmitted in any form or by any means, electronic, mechanical, photocopying, recording, or otherwise, without written permission from the publisher.

For more information, write to Bearport Publishing, 5357 Penn Avenue South, Minneapolis, MN 55419. Printed in the United States of America.

Contents

Seeing with the Brain 4
Is It This or That? 6
What Kind of Baby? 8
Confusing Cubes 10
Look at the Columns, Man! 12
Is It Really There? 14
Let's Face It! 16
Scary Skulls 18
Which Way Is Up? 20

Make Your Own Changing Cube Illusion . . 22
Glossary 23
Index 24
Read More 24
Learn More Online 24
About the Author 24

Seeing with the Brain

The world is full of amazing things to see. But how do you know what you're looking at? Your eyes and brain work together to tell you.

When you look at an object, light bounces off the object and comes to your eyes. Then, **nerves** send signals to your brain, which uses your memories from other **experiences** to figure out what you're seeing in the moment. You **recognize** the object!

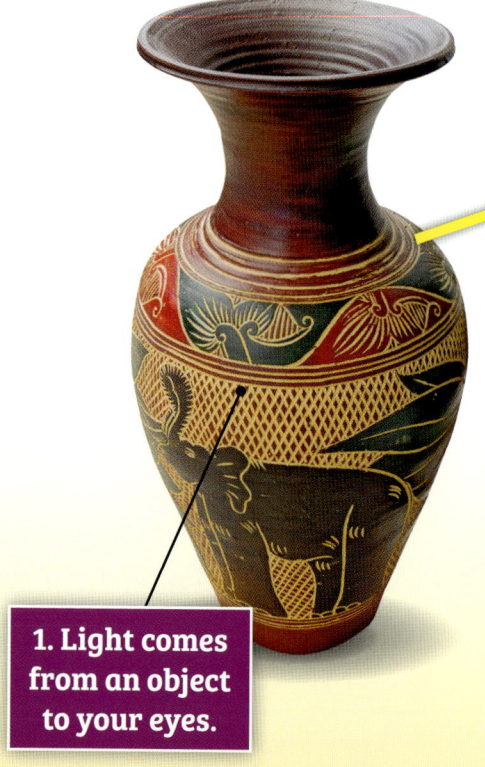

Your brain is very good at recognizing objects, even when they appear in different ways. An apple may look big or small. It may be sitting in the dark or on its side. But you can still recognize it as an apple.

1. Light comes from an object to your eyes.

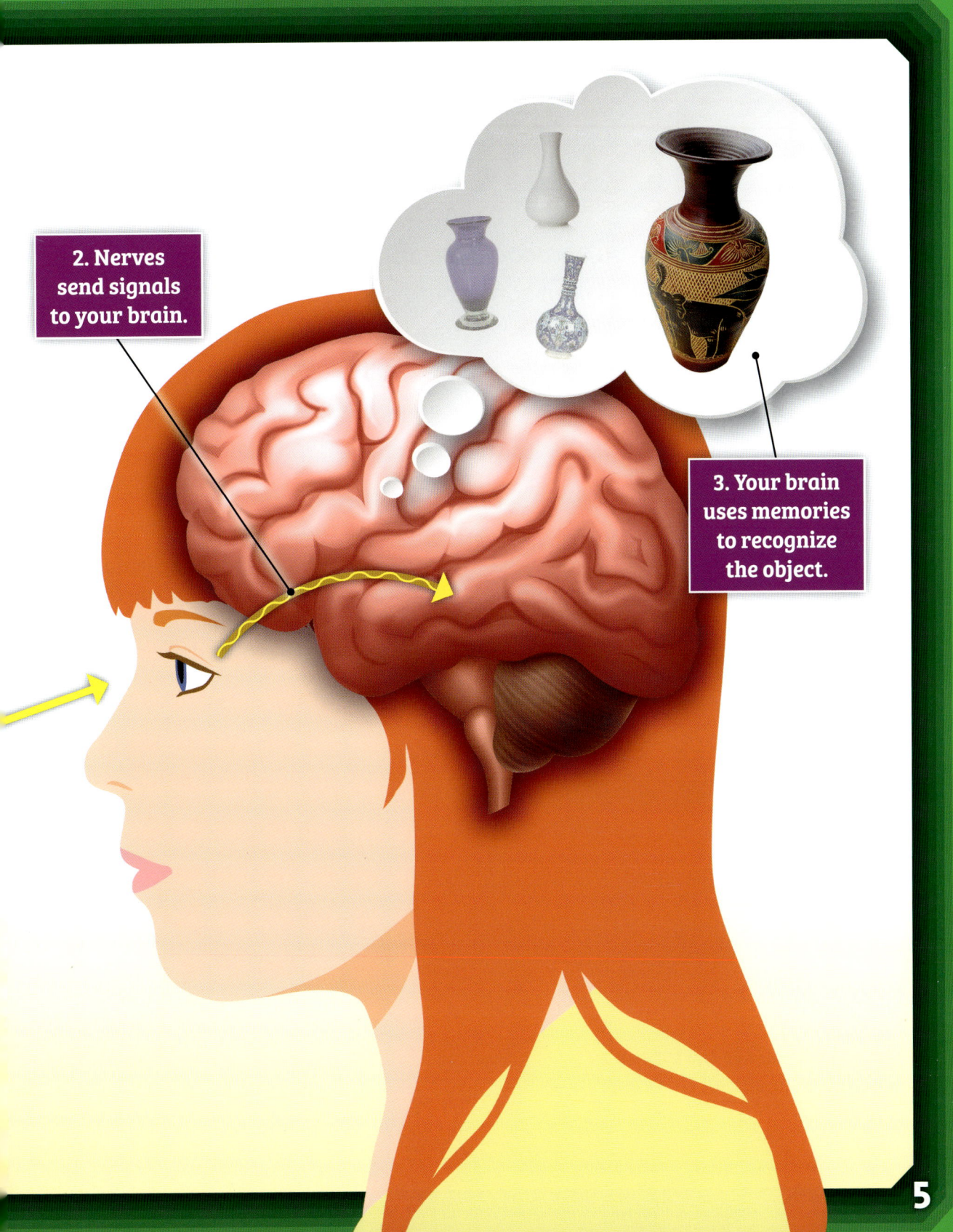

Is It This or That?

Your brain usually recognizes objects very quickly. But what happens when something suddenly looks like something else? Some pictures can make an object seem to be two things at once! Let's explore some of these **ambiguous** pictures. Enjoy the **optical** illusions, and have fun trying to decide what it is that you're *really* seeing.

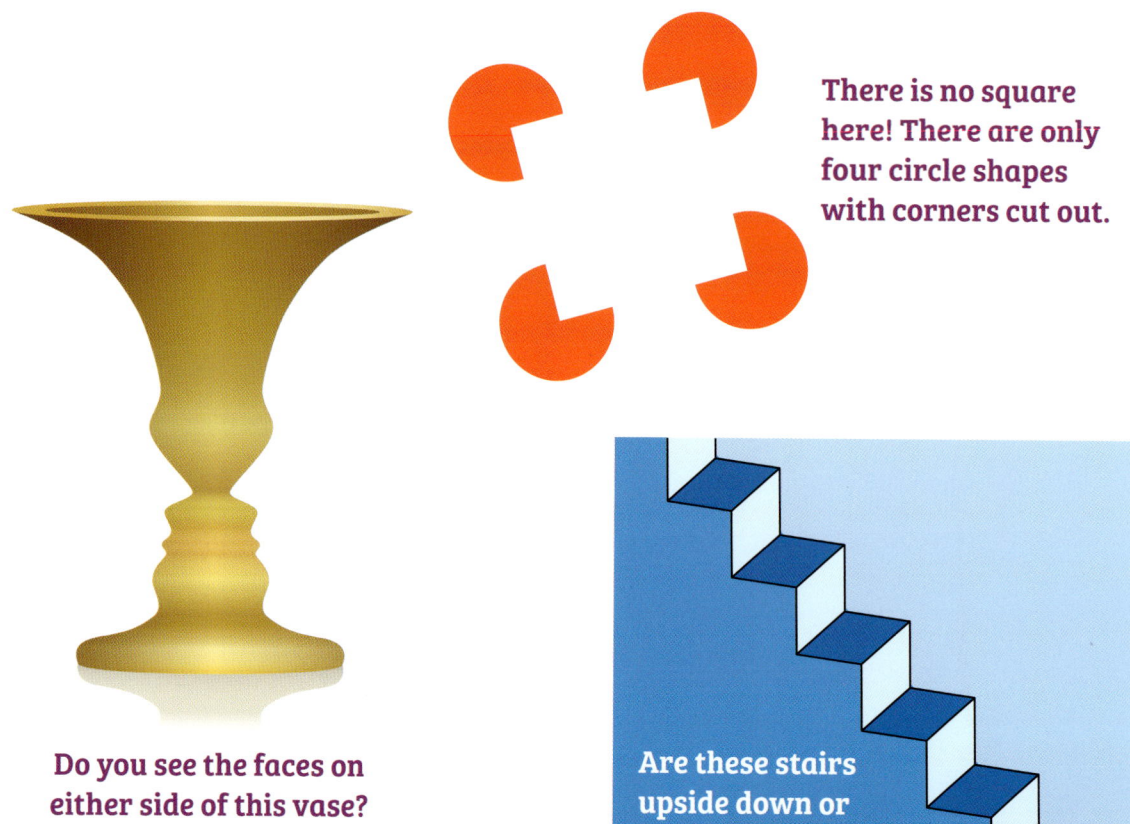

There is no square here! There are only four circle shapes with corners cut out.

Do you see the faces on either side of this vase?

Are these stairs upside down or right side up?

This famous ambiguous drawing was made more than 100 years ago. Do you see a young woman turning her head away? Or maybe it looks like an old woman with a large nose.

What Kind of Baby?

Everybody loves baby animals. Isn't that a cute baby on the grass? But what kind of animal is it? Is it a bunny, or is it a duckling? Well . . . it's both!

The first duck-rabbit illusion may have been drawn in 1892. Since then, the picture has appeared in many different forms. Some brain scientists even use this illusion to study how people understand the world around them.

Confusing Cubes

Patterned shapes can make especially tricky illusions. Look at the orange, red, and yellow pattern. How many **cubes** do you see? Are there six, or are there seven? If you shift your eyes, do different cubes seem to form?

The cube illusion is used on many things, from floors to quilts. The simple pattern of lighter and darker colors gives the illusion of **depth**. If you look closely, these cubes can change, too!

An old quilt

A wooden floor

The number of cubes you see depends on how you look at the picture. Try looking for seven cubes with red tops. Then, try looking for six cubes with red bottoms.

Look at the Columns, Man!

When you look at the blue columns, do you notice anything about the space between them? Does your brain see anything it recognizes in the gaps?

The empty space between objects is often called **negative space**. The negative space between these columns is shaped like a man.

Is It Really There?

Negative space can be tricky. Your brain might use the space between objects to see a shape. But is it *really* there?

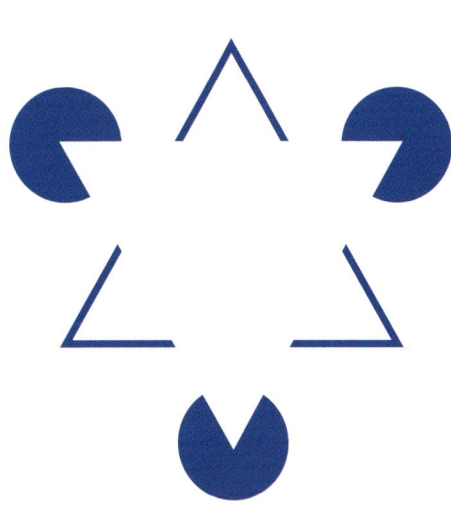

How many triangles do you see? There are none.

The red shapes in this picture give the illusion of a white **sphere**. But it's not there.

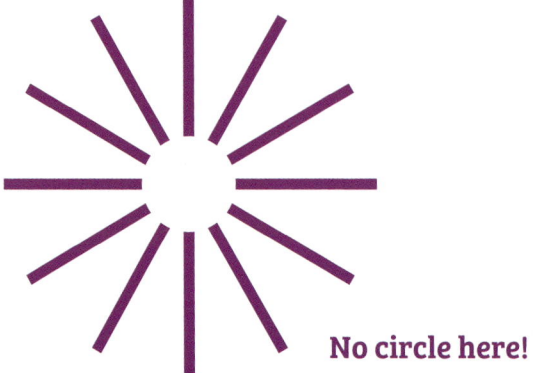

No circle here!

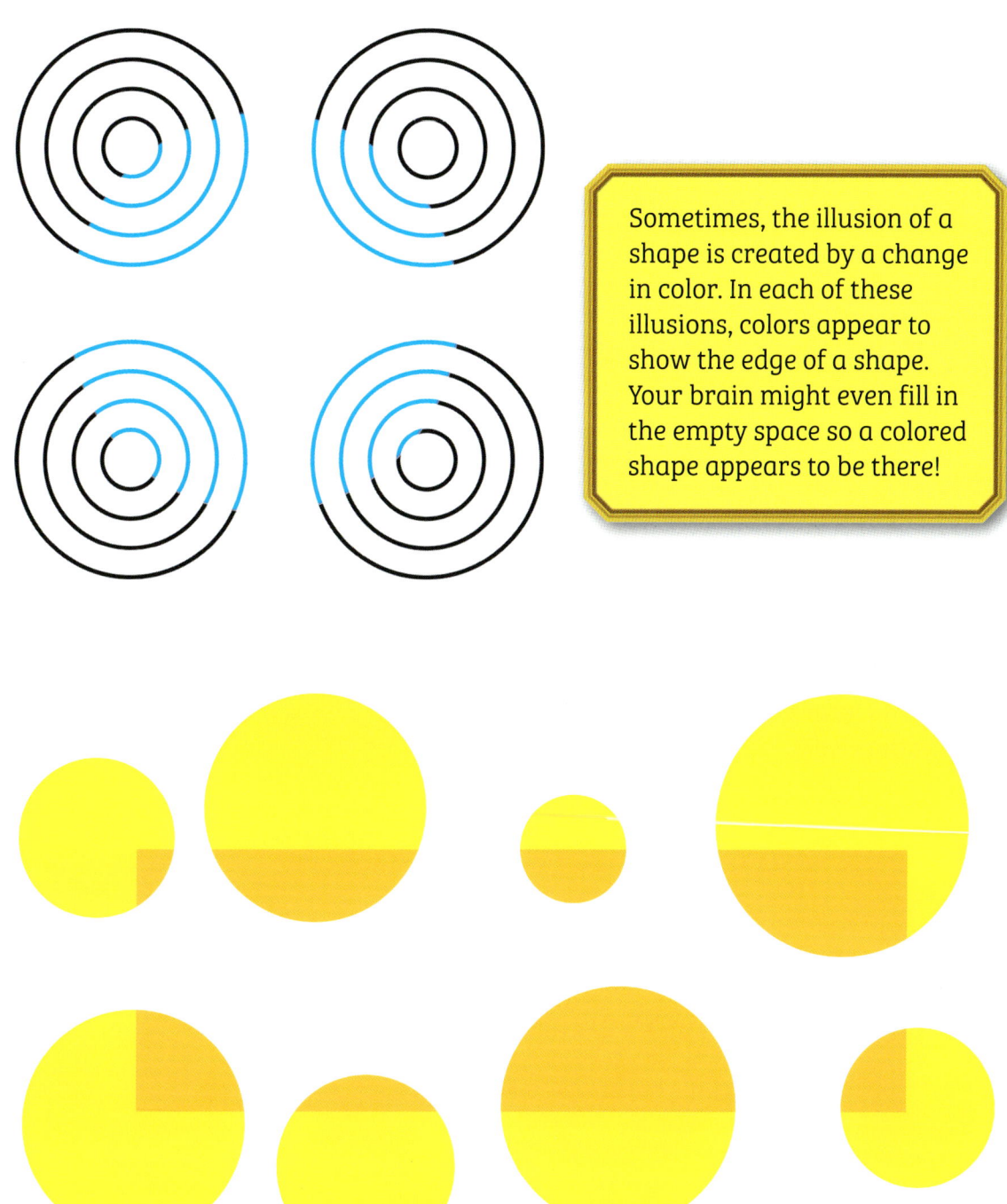

Sometimes, the illusion of a shape is created by a change in color. In each of these illusions, colors appear to show the edge of a shape. Your brain might even fill in the empty space so a colored shape appears to be there!

Let's Face It!

People are very good at recognizing human faces. This is such a big part of how your brain works that you can even see faces in things that aren't alive. What's even more surprising is that you often recognize **emotions** in those faces!

Which emotions do think of when you look at these pictures? Do they look happy, sad, scared, or something else?

Artists sometimes use different objects to create faces in fun and interesting ways. This painting of a Roman leader is more than 400 years old. Do you recognize the objects used to make his face?

Scary Skulls

Seeing skulls in pictures is similar to seeing faces—it's just a bit creepier. For example, the famous 1892 drawing, *All Is Vanity*, shows a woman sitting at a table, looking in the mirror. But if you move back, you can see something else . . . a skull!

All Is Vanity by Charles Allan Gilbert

The skull-like coloring on the back of this moth is scary enough. To make it worse, the insect will make loud chirping noise if bothered!

18

This painting of two men was created in 1533. But what is that strange shape at the bottom? Move close to the right side of the painting about half way up, and then look at the shape from that position. Do you see what the artist has hidden?

Which Way Is Up?

Look at the faces in the girl's school photo to the right. It's easy to see that something is wrong with the eyes and mouth of the face in the lower right corner. Interestingly, it's not as easy to see the same thing when the picture is upside down. Look at the top right picture, and then turn the book upside down to look at it again.

Artists sometimes create ambiguous pictures that can be viewed both upside down and right side up.

This drawing shows that words can sometimes be seen two different ways. You've seen a lot of puzzles in this book. Turn the book upside down and read the new message.

The bottom right picture looks very wrong, but the the other three look fine. What happens when you turn the book upside down?

21

— MAKE YOUR OWN —
Changing Cube Illusion

Make your own cube optical illusion. Amaze your family and friends when one side of the cube appears to switch from front to back!

YOU WILL NEED:
- A sheet of graph paper
- Yellow and black markers

1. Starting slightly off center on the graph paper, color in a square with the yellow marker. Make it 10 spaces wide and 10 spaces tall.

2. Next, use your black marker to draw a line around the square.

3. Then, start at the top right corner and count five spaces down and five spaces to the right. Use this point as the upper right corner and draw another square the same size.

4. Draw a straight line from the upper right corner of one square to the upper right corner of the other. Repeat with the remaining three corners.

5. Now, ask your family and friends to look at your illusion. Is the yellow side in front or in back? Can they see it both ways?

Glossary

ambiguous easily understood in more than one way

cubes objects that have six square sides

depth the appearance of being three-dimensional

emotions feelings

experiences things that have happened in the past

negative space the empty space between objects

nerves tiny parts of the body that carry messages to and from the brain

optical related to sight or seeing

recognize to understand or be familiar with

sphere a ball-shaped object

Trompe l'oeil means to fool the eye. This style of painting makes a flat surface appear to have depth.

The wall around this window is flat.

Index

artists 17, 19–20
brain 4–6, 8, 12, 14–16
circles 6, 14
colors 10, 15, 22
cubes 10–11, 22
depth 10, 23
eyes 4, 20, 23
faces 6, 16–18, 20
negative space 12, 14
nerves 4–5
skulls 18
triangles 14

Read More

Claybourne, Anna. *Puzzling Pictures (The Science of Optical Illusions).* New York: Gareth Stevens Publishing, 2020.

Felix, Rebecca. *Optical Illusions to Trick the Eye (Super Simple Magic and Illusions).* Minneapolis: Abdo Publishing, 2020.

Learn More Online

1. Go to **www.factsurfer.com** or scan the QR code below.
2. Enter "**Faces and Vases**" into the search box.
3. Click on the cover of this book to see a list of websites.

About the Author

Spencer Brinker lives and works in Minnesota. Having twin daughters sometimes makes it seem like he's seeing double!